PICTURE LIBRARY
LASERS
AND HOLOGRAMS

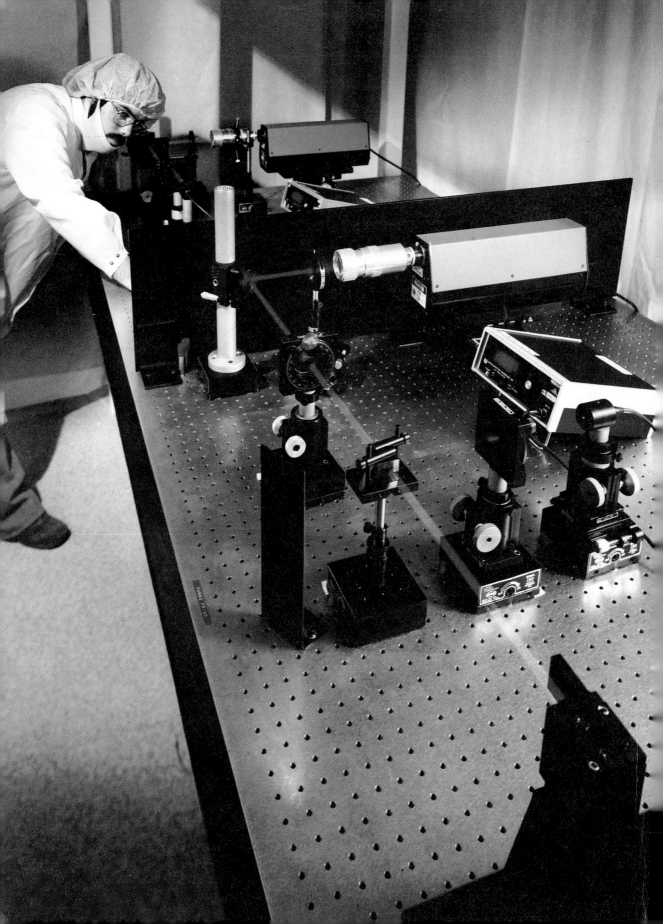

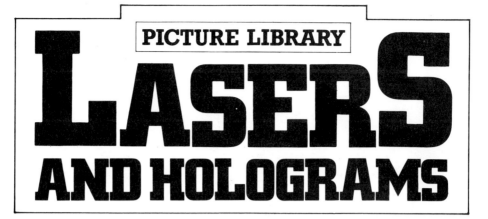

PICTURE LIBRARY
LASERS AND HOLOGRAMS

N. S. Barrett

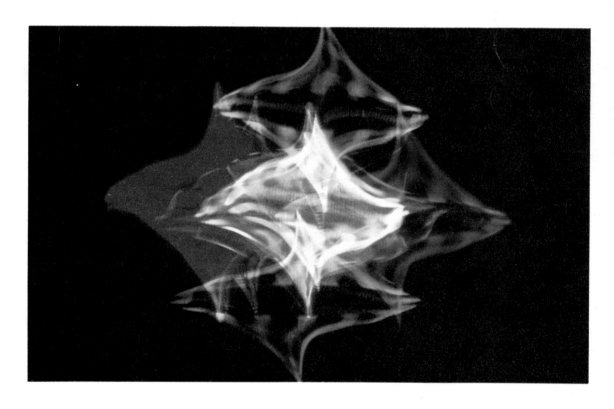

Franklin Watts

London   New York   Sydney   Toronto

© 1985 Franklin Watts Ltd

First published in Great Britain
   1985 by
Franklin Watts Ltd
12a Golden Square
London W1

First published in the USA by
Franklin Watts Inc
387 Park Avenue South
New York
N.Y. 10016

First published in Australia by
Franklin Watts
1 Campbell Street
Artarmon, NSW 2064

UK ISBN: 0 86313 227 8
US ISBN: 0-531-04946-9
Library of Congress Catalog Card
Number: 84-52000

Printed in Italy

Designed by
Barrett & Willard

Photographs by
Barclays Bank
Bell Laboratories
Bergström
Coherent (UK)
Control Laser
EMI Records
Ferranti
Hughes Aircraft Co
Lasergage
Laserpoint
Laserium/London Planetarium
Lawrence Livermore Laboratory
Light Fantastic
Mowlem & Co
NASA
NCR
Philips LaserVision
Rutherford & Appleton Laboratories
Sperry
The Who Group
ZEFA/Photri

Illustrated by
Mike Saunders

Technical Consultant
William Burroughs

Contents

Introduction	6
Making a hologram	8
Uses of lasers and holograms	10
Lasers at play	16
Measuring with lasers	20
Sending signals	22
Lasers at war	24
Holograms	26
The story of lasers	28
Facts and records	30
Glossary	31
Index	32

Introduction

△ Using a laser beam to cut steel. Lasers are also used to cut soft materials, such as cloth or canvas. The heat of the beam seals the cut edges and avoids fraying.

Lasers produce very strong beams of light. A laser beam is light of one color. It is thin and narrow, unlike white light, which spreads out. Laser light is brighter than the Sun's light.

Holograms are three-dimensional, or "solid," photographs. They are produced by focusing laser beams on a special photographic plate.

Lasers are powerful enough to cut through thick steel, yet they are also used in delicate eye surgery. Laser light is used for making very accurate measurements and for producing special effects at pop concerts.

Holograms are used in engineering and for storing information.

△ A hologram of a cup and saucer. The image is so real that you feel you can reach out and touch it. But there is nothing solid there.

Making a hologram

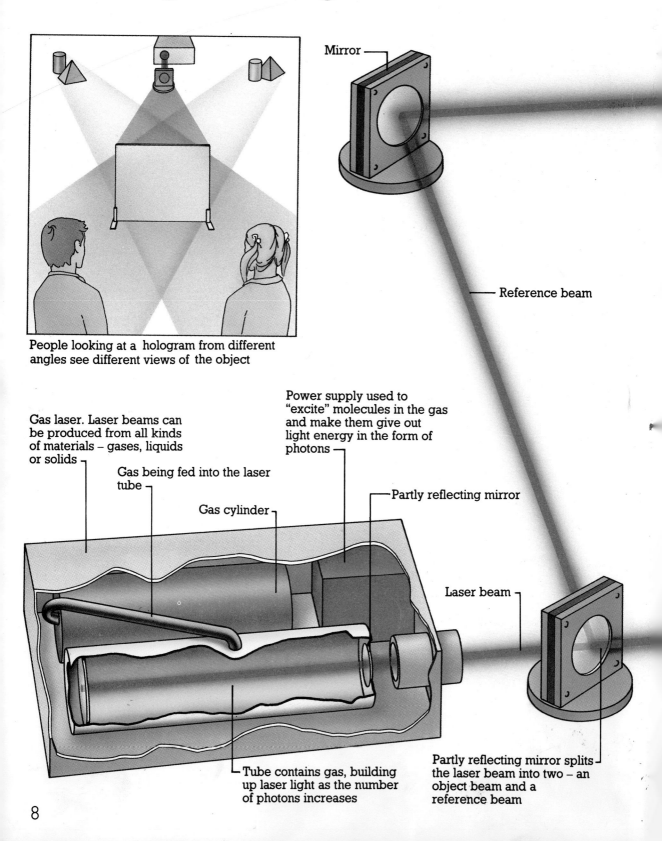

People looking at a hologram from different angles see different views of the object

Gas laser. Laser beams can be produced from all kinds of materials – gases, liquids or solids

Gas being fed into the laser tube

Gas cylinder

Power supply used to "excite" molecules in the gas and make them give out light energy in the form of photons

Partly reflecting mirror

Tube contains gas, building up laser light as the number of photons increases

Partly reflecting mirror splits the laser beam into two – an object beam and a reference beam

Laser beam

Mirror

Reference beam

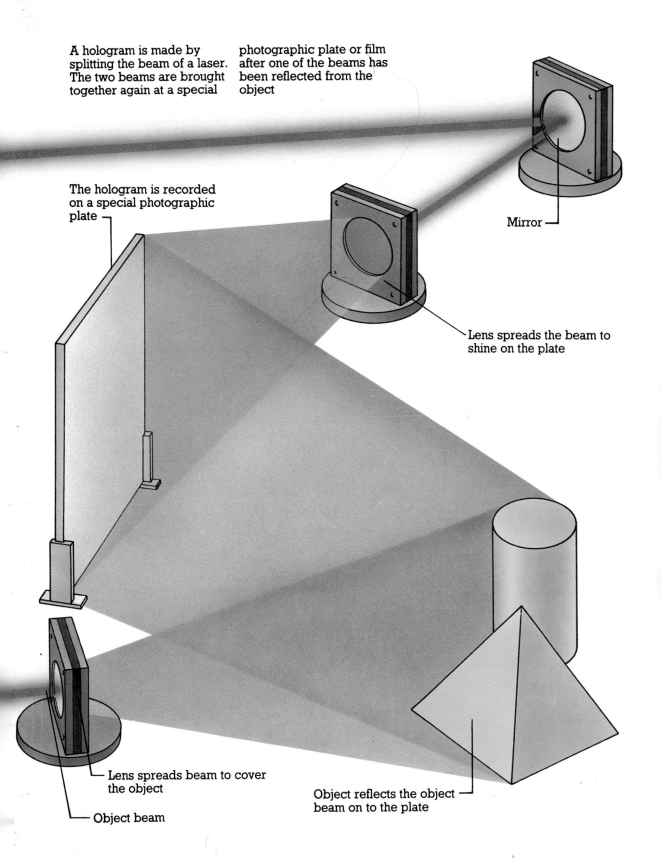

Uses of lasers and holograms

Laser beams are used for cutting and drilling materials and for welding, or joining metal by heat.

Lasers are excellent tools for all of these tasks. They do not touch the material, so they do not wear out or break. Lasers are very fast and accurate, and, by use of mirrors, they can get to places that ordinary tools cannot reach.

△ Using a laser for welding metals together. The power of a laser beam is easy to control and produces better welds than methods using a flame.

Light travels in straight lines, and laser light produces a narrow beam. Such a beam makes a perfect straight line for aligning things, especially over long distances.

This property makes lasers a simple and ideal tool for use in construction work. They are used for alignment in building bridges, roads, tunnels and other structures.

▽ Lasers are used in construction work. This beam checks the alignment of a tunnel.

Lasers are used for storing information, for printing and for reading television video discs.

Special lasers can store sounds and pictures on laser discs. These are stronger than ordinary magnetic tapes and discs and are less likely to be damaged. Video discs are also tough. They never wear out because the laser beam that plays them is too weak to do any damage.

△ The turntable of a video disc player. The scanning lens tracks outward from the center of the disc during play. A laser beam shines on to the underside of the disc and is reflected back through the lens to produce the pictures.

In printing, lasers can be used to engrave the printing plates or to set type photographically.

The bar codes on supermarket goods and in library books are read by lasers. A bar code is computer information in the form of light and dark lines of different thicknesses. Laser beams from a scanner or from special laser wands read the information and send it to a computer.

▽ In a supermarket, a laser beam reads the bar code as the product is moved across a special scanner.

There are many uses for lasers in medicine. Surgeons use them in delicate operations, especially on the eyes. They can aim them with pin-point accuracy, and use laser energy to burn away diseased parts or repair damaged areas.

Lasers are used in similar ways in dentistry. Holograms are used for recording information about patients' teeth.

△ Holograms of teeth provide a simple way of storing information without the need to keep bulky molds.

▷ Laser beams become a delicate tool in the hands of an eye surgeon. In most operations with lasers, there is no need for an anesthetic as the patient suffers no pain.

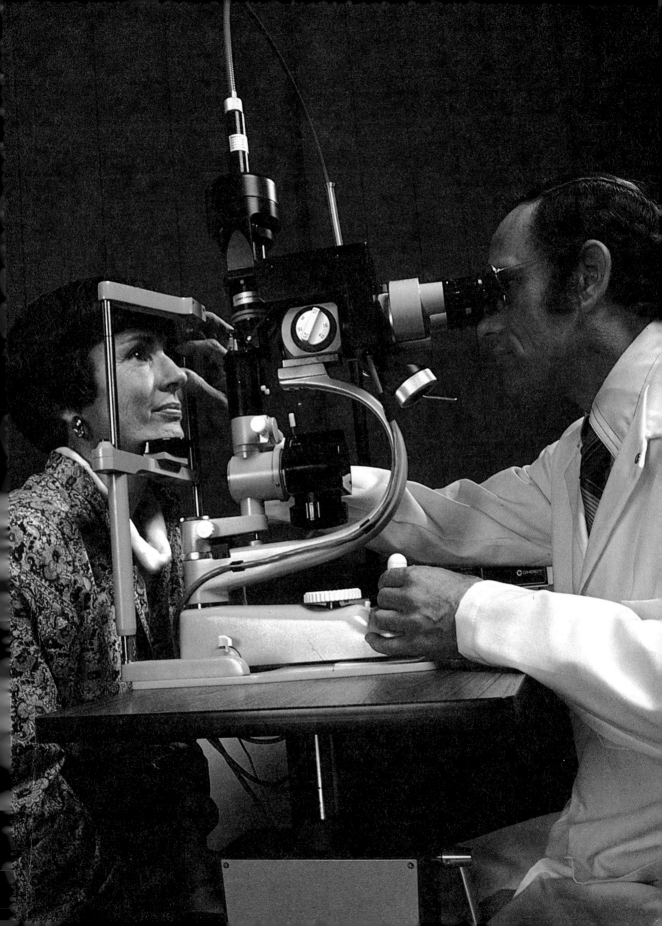

Lasers at play

The color and excitement of laser light are qualities that have made it spectacular entertainment. Laser beams can be controlled by the beat of the music and are often used by pop groups and in discos.

Musical laser shows are an entertainment in themselves. Brilliant, colorful dancing patterns are projected on to walls and ceilings in time to music of all kinds.

▷ Laser light shows are a feature of many science museums. Beautiful patterns are projected to music.

▽ Laser light provides an interesting background for a promotional photograph of a pop group. Lasers are often used on stage by rock bands to add color and excitement to their performances.

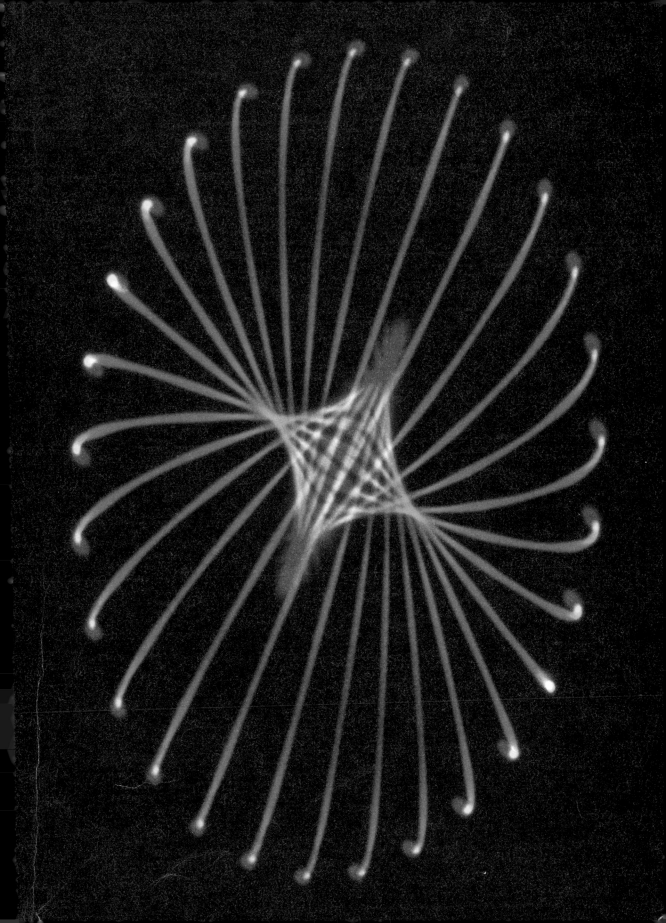

▷ Laser displays are popular in clubs and discos. The most common type of equipment for this kind of work uses gas or dye lasers to produce the distinctive colors. The "fingers of light" display in the picture was produced by a krypton laser. Krypton is a gas.

Measuring with lasers

Lasers have special properties that make them ideal for measuring distance. A laser beam can be bounced back off a distant object to provide a very accurate measurement of its distance.

At some big airports, lasers are used to measure the exact height of clouds.

▷ A laser beam reflected from the Moon provides an accurate measurement of the distance between Moon and Earth.

▽ Laser reflectors were set up on the Moon by astronauts.

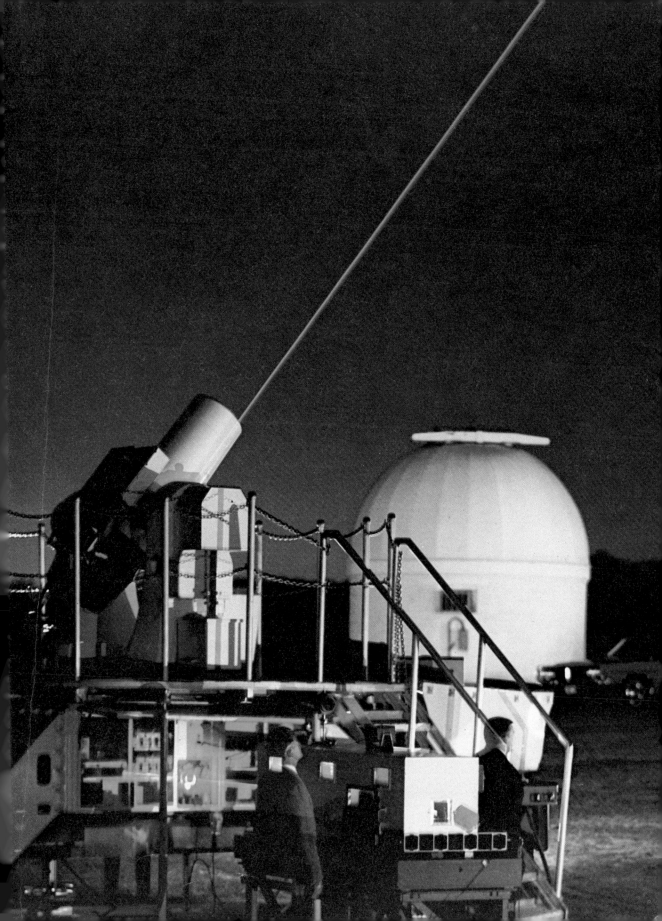

Sending signals

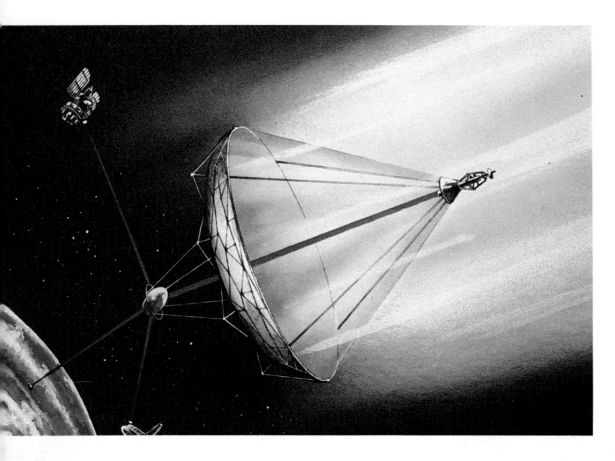

△ An illustration showing how the rays of the Sun are used to work a laser on a satellite. The beam may be used for sending messages to other satellites or to ground stations.

Laser beams can be used for sending sound and picture signals. The beams do not spread out like radio waves, so they are stronger and go directly to a receiver with little chance of interception.

This works well in space, where nothing interferes with the beam. But in the atmosphere, clouds and fog sometimes obstruct laser signals.

Scientists have found a very efficient way of using laser light to carry information. They send it through hair-thin glass cables called fiber optics.

Some telephone systems have already begun using fiber optics. They are much thinner than ordinary copper cables, yet can carry many more calls.

▽ Light sent through fiber optics is "bent" when the fiber bends. It is guided along the inside walls of these fine glass rods. Very little laser light is lost as it passes through. As a result, in a telephone or cable TV system, the signals need less boosting than with other methods of transmission.

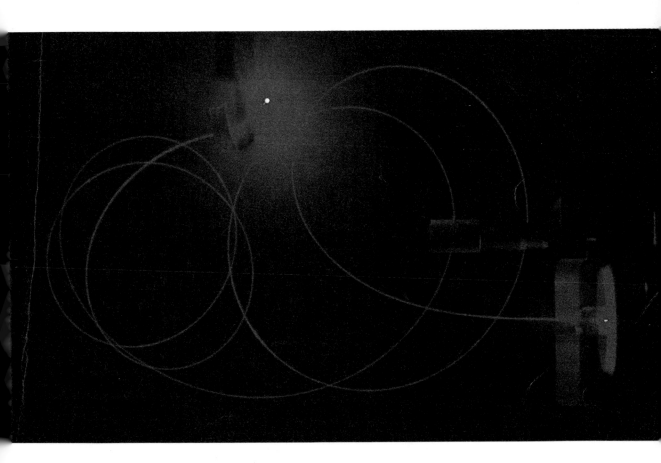

Lasers at war

Lasers have many military uses. They are used for range-finding and for guiding weapons on to targets on land, at sea and in the air.

Range-finding involves bouncing an invisible laser beam off a target to calculate the exact distance and sometimes the speed. A laser beam may also be used to guide a missile directly on to a target.

△ A soldier uses laser range-finder binoculars. A timing device measures the time the laser beam takes to return from the target. The distance of the target is calculated automatically and is displayed on one of the eyepieces.

△ Laser equipment is mounted in the nose cone of a Harrier jet. Combat planes use lasers for guidance and for target-seeking.

Ray guns are popular weapons in science fiction films. Large lasers are needed to produce such powerful beams, so hand-held laser weapons are not yet possible.

But much research is being carried out to produce laser weapons as defense against long-range missiles. These would be used in planes or satellites.

Holograms

Holograms are made by splitting a laser beam into two, as shown on pages 8–9. The two beams come together again after one has been reflected from the object.

A hologram is not just a three-dimensional photograph. You can see different views of the object as you change your angle of viewing.

▽ Even though this hologram has a ghostly appearance, the telephone still looks real enough to use.

The holographic plate or film is not itself a picture. The three-dimensional image is produced by shining light on it. For the best effect, the same laser light that made the plate should be used.

Simple holograms can be produced on special plastic and are used for such things as book and record covers. In engineering, holograms are used for testing that parts have been made accurately.

△ Holograms are being used on credit cards to make them difficult to forge. The hologram is embedded in the plastic.

The story of lasers

First the maser
In the early 1950s, scientists developed an electronic device which they called a maser. It is used to amplify, or strengthen, radio and light waves.

The idea of the laser came a few years later, and scientists in both the US and the USSR were responsible for its invention.

△ An early laser experiment at the Hughes Aircraft laboratories.

Naming the invention
The term laser is made up from the initials of **L**ight **A**mplification by **S**timulated **E**mission of **R**adiation. This describes how a laser works. The atoms of the material being used to produce the laser light are stimulated, or excited into great movement, by an energy source such as an electric current. As a result, the atoms emit, or give off, tiny "packets" of light energy called photons. These hit other excited atoms, and more photons are given off. This is called stimulated emission of radiation. The overall effect is an amplification, or strengthening, of the beam of light.

The first laser
The first workable laser was built by Theodore Maiman and other American scientists in California in 1960. The material they used was a man-made ruby rod. A simple flash tube was coiled around the rod, and powerful flashes of light beamed at it. The result was pulses of red laser light.

New materials
Scientists soon began to produce continuous laser light and to use different kinds of materials. They started to use liquids and gases as well as solid crystals, such as ruby and diamond.

△ An experiment on a ruby laser in the Sperry laboratories, New York.

△ A hologram reconstructed by the use of ordinary white light.

A new kind of photography

A kind of hologram was produced as far back as 1948 by a British scientist, Dennis Gabor. But he did not have the equipment to develop his invention.

Holography became possible only with the invention of lasers. At first, a hologram could be reconstructed only by using the same laser light that had created it in the first place. Now, ordinary white light can be used.

△ Laser light traveling through an optical fiber.

Finding uses

The invention of the laser caused excitement in the scientific world. But it was not until the 1970s that lasers began to come into everyday use. Then all kinds of uses were found, in many different areas.

More and more new uses are still being found to take advantage of the laser's remarkable properties – power, beauty and precision.

△ Scientists continue to explore new ways of using the power of lasers.

We see plenty of examples of the beauty of lasers in art and entertainment. Its accuracy is evident in surgery and measurement. We have seen how lasers can be used with fiber optics to run telephone and other communications systems. But scientists have a long way to go before they tap the enormous power that can be generated by means of lasers.

Facts and records

The Shiva laser

Scientists are using lasers in an attempt to harness the mighty power of the atom. In the Sun, enormous energy is produced when hydrogen changes to helium, another gas. This is called fusion, and needs very special conditions and temperatures of millions of degrees.

A giant laser has been built in California, at the Lawrence Livermore Laboratory, for experiments on fusion. It produces more power than all the world's power stations put together – but only for the tiniest part of a second.

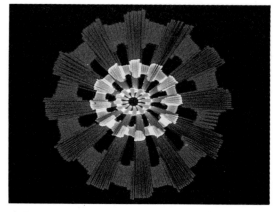

△ Laser art – the projection of laser images in a laser show.

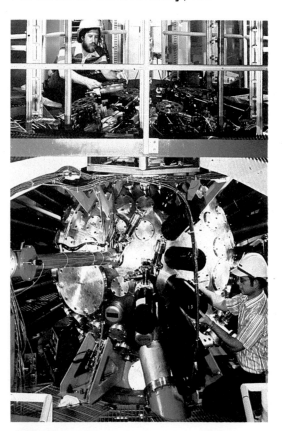

△ Engineers work on some of the complex machinery of the Shiva laser.

Laser shows

In laser shows, different color lasers are deflected by mirrors inside a projection unit to produce brilliant patterns and shapes. These shapes seem to dance to the music, which is linked with the projector. Giving these shows is a new form of art.

Broken plates

If you break a holographic plate, the hologram is not destroyed. Each piece of the plate will contain a hologram of the whole object, but fainter than on the complete plate.

Glossary

Bar code
The pattern of light and dark lines used on products sold in shops and in library books, for example. The code carries computerized information which can be read by laser scanning.

Credit card
A plastic card that is used by people to buy goods or pay bills without using cash or checks. To make them almost impossible to forge, they are now being produced with holograms inside the plastic.

Fiber optics
Very fine flexible glass rods used for "carrying" light.

Fusion
The process that takes place when two atoms of hydrogen are "fused" together to make an atom of helium. Scientists hope to produce this reaction in a controlled way by using very powerful lasers. It could provide us with an unlimited source of power.

Object beam
The part of the split laser beam that is shone on to the object in the making of a hologram.

Photons
Small "packets" of radiation that form waves such as beams of light.

Range-finder
An instrument used to find the range of an object – that is, how far away it is.

Ruby
The first substance used for making lasers. Ruby is a precious stone that can be made artificially. It contains substances well suited for the production of laser light.

Scanner
A machine that can "read" computerized or other encoded information by means of reflected light. Laser scanners are used on bar codes and video discs, for example.

Video disc
A disc that can be played to produce pictures and sound on a video player.

Welding
Joining materials together by heating. Lasers are used for welding metal. Special lasers also weld parts of the eye.

Index

alignment by laser 11
astronaut 20
atmosphere 22

bar code 13, 31

cable TV 23
computer 13
construction 11
credit card 27, 31
cutting 6, 7, 10

dentistry 14
diamond 28
drilling 10
dye laser 18

engineering 27
eye surgery 14, 15

fiber optics 23, 29, 31
fusion 30, 31

Gabor, Dennis 29
gas laser 8, 18, 28
guided weapons 24

hologram 6, 7, 8, 9, 14, 26, 27, 29, 30, 31
holographic plate 9, 27, 30
holography 29

krypton laser 18

laser art 30
laser display 17, 18, 19

laser power 29
laser reflector 20
light show 16, 17, 30

Maiman, Theodore 28
maser 28
measuring 20
medicine 14
military laser 24, 25
Moon 20, 21
musical laser show 16

object beam 8
operation 14, 15

photons 8, 28, 31
printing 13

radio waves 22, 28
range-finding 24, 31
ray gun 25
reference beam 9
ruby 28, 31

satellite 22, 25
scanner 12, 13, 31
Shiva laser 30
space 22
Sun 6, 22, 30

telephone 23

video disc 12, 31

welding 10, 31
white light 6, 29

DATE DUE

	AP 14'90	JE 04 92	
JA 28	MY 12 90	FE 26 95	
JR 5'97	SE 03		
AP 28 8	NO		
MY 20'8	DE 05 91		
JE 10'8	FE 26 91		
SE 5'8	MY 15 92		
AP 19			
JY 14 89	29 93		
NO 6'8	MY 02 9		
FE 9'90	MY		
	OC 10 95		

```
J                          j112623
535.5
BAR
Barrett, N. S.
    Lasers and holograms
```

MARION CARNEGIE LIBRARY
1298 7th Avenue
Marion, Iowa 52302